Collins

easy le____g

Fractions and decimals

Ages 7-9

$$\frac{3}{4} = 0.75$$

Adam Blackwood
Melissa Blackwood

How to use this book

- Find a quiet, comfortable place to work, away from other distractions.
- Tackle one topic at a time.
- Help with reading the instructions where necessary, and ensure that your child understands what to do.
- Help and encourage your child to check their own answers as they complete each activity.
- Discuss with your child what they have learnt.
- Let your child return to their favourite pages once they have been completed, to talk about the activities.
- Reward your child with plenty of praise and encouragement.

Special features

- Yellow boxes: Introduce and outline the key ideas in each section.
- Orange shaded boxes: Offer advice to parents on how to consolidate your child's understanding.

Published by Collins
An imprint of HarperCollins*Publishers*
1 London Bridge Street
London SE1 9GF

Browse the complete Collins catalogue at www.collins.co.uk

© HarperCollins*Publishers* 2015

10 9 8 7

ISBN 978-0-00-813445-7

The authors assert their moral right to be identified as the authors of this work.

British Library Cataloguing in Publication Data

A Catalogue record for this publication is available from the British Library

Written by Adam Blackwood and Melissa Blackwood
Page design by QBS Learning
Cover design by Sarah Duxbury and Paul Oates
Project managed by Andy Slater
Printed and bound in Great Britain by Bell and Bain Ltd, Glasgow

Contents

What are fractions and decimals?

A **fraction** is part of a whole.

$\dfrac{4}{6}$ ← The top number is called the **numerator**.
← The bottom number is called the denominator.

A **unit** fraction has a numerator of one. $\dfrac{1}{2}, \dfrac{1}{8}$

A **non-unit** fraction has a numerator greater than one,
but smaller than the denominator. $\dfrac{2}{5}, \dfrac{3}{4}, \dfrac{5}{8}$

Decimal numbers contain a decimal point.
The decimal point separates the units from the tenths column.

tens	units	.	tenths
2	4	.	7

Decimals are numbers that are between whole numbers.
For example, 0.5 is half way between 0 and 1

0   0.5   1

---

1 Colour the fractions red and the decimals blue.

$\dfrac{3}{4}$ 12.3   6.8 $\dfrac{8}{10}$ 19.8 $\dfrac{1}{9}$ 31.5 $\dfrac{1}{7}$

---

2 Write these number words as decimal numbers. The first one has been done for you.

four units, two tenths — 4.2   three units, five tenths — ☐

six tens, one unit, nine tenths — ☐   five tens , five units, eight tenths — ☐

seven tens, zero units, five tenths — ☐   two tens, eight units, six tenths — ☐

3 Write these decimal numbers as number words. The first one has been done for you.

24.5   Two tens, four units, five tenths

8.6   _____

70.7   _____

16.2   _____

4 Write these fractions using numbers, then circle unit fraction or non-unit fraction. The first one has been done for you.

one-tenth $\frac{1}{10}$ (unit fraction)   non-unit fraction

two-fifths   [ ]   unit fraction   non-unit fraction

four-twelfths [ ]   unit fraction   non-unit fraction

seven-eighths [ ]   unit fraction   non-unit fraction

one-half   [ ]   unit fraction   non-unit fraction

5 Complete the numerators to match the labels.

$\frac{\Box}{7}$ unit fraction $\frac{\Box}{8}$ unit fraction

$\frac{\Box}{7}$ non-unit fraction $\frac{\Box}{6}$ non-unit fraction

$\frac{\Box}{5}$ non-unit fraction $\frac{\Box}{5}$ unit fraction

Children will use mathematical vocabulary such as numerator and denominator at school. Try to use the correct terminology used in this book when talking to your child.

Tenths

You will be able to recognise a half and a quarter quite easily. A **tenth** is a smaller fraction. One whole circle divided into tenths looks like this.

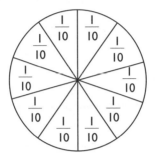

Each part is one-tenth. There are ten-tenths. They are all the same size.

You can also write one-tenth as a decimal.

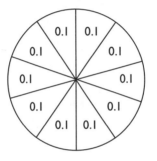

Each part is 0.1 There are ten of these parts. They are all the same size.

Dividing into tenths is the same as dividing a number by 10.

We can use a place value table to see how a number can be divided by 10.

$$1 \div 10 = 0.1$$

Units	·	tenths
1	·	0
0	·	1

If you divide a units number by 10, the number moves one place to the right.

---

1 Look at the number lines and fill in the gaps.

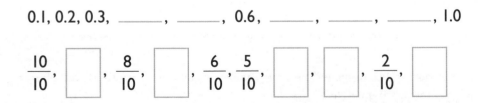

0.1, 0.2, 0.3, _____, _____, 0.6, _____, _____, _____, 1.0

$\frac{10}{10}$, ☐, $\frac{8}{10}$, ☐, $\frac{6}{10}$, $\frac{5}{10}$, ☐, ☐, $\frac{2}{10}$, ☐

2 Draw lines between the fractions and decimal numbers that have the same value.

$\frac{4}{10}$ $\frac{3}{10}$ 0.8 $\frac{7}{10}$ 0.9

0.3   0.1

$\frac{1}{10}$ 0.7 $\frac{9}{10}$ 0.4

Which number is the odd one out? _____

3 Colour in the correct number of objects in each picture to match the label.

0.1 or $\frac{1}{10}$ of the tokens in the money box   0.6 or $\frac{6}{10}$ of the bag of sweets

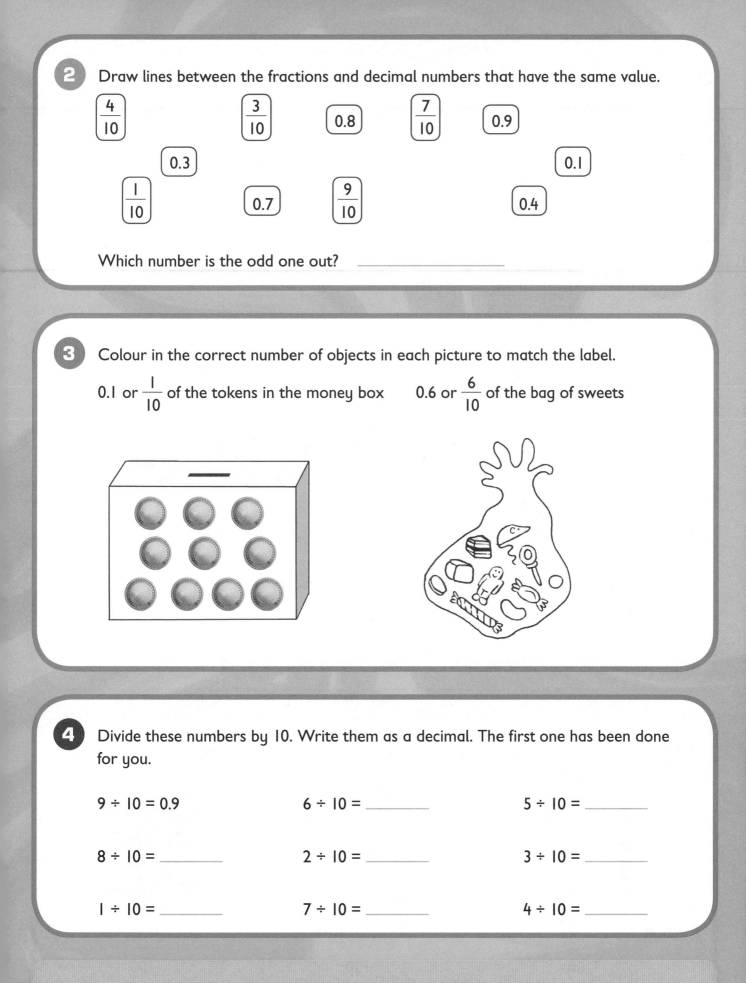

4 Divide these numbers by 10. Write them as a decimal. The first one has been done for you.

9 ÷ 10 = 0.9   6 ÷ 10 = _____   5 ÷ 10 = _____

8 ÷ 10 = _____   2 ÷ 10 = _____   3 ÷ 10 = _____

1 ÷ 10 = _____   7 ÷ 10 = _____   4 ÷ 10 = _____

It is important your child understands the link between tenths written as a fraction and as a decimal. It is the same number, just written in two different ways.

Hundredths

A **hundredth** is one part of a whole that has been divided into tenths and then each tenth divided into tenths again.

One hundredth can be written as a fraction like this

$$\frac{1}{100}$$

One hundredth can be written as a decimal like this

0.01

We can count in hundredths. A number line from 0 to 1 in hundredths would look like this:

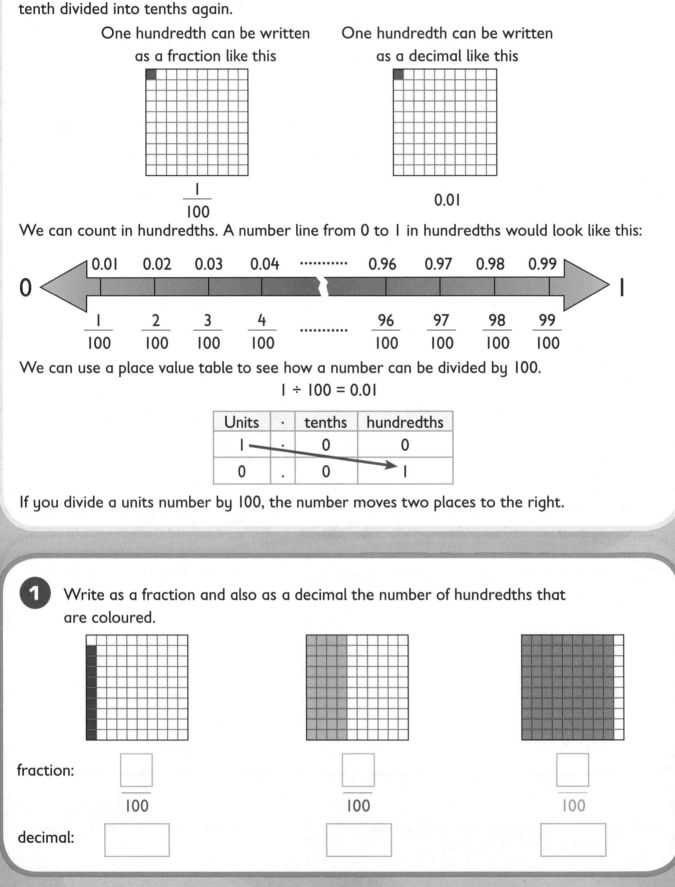

| 0.01 | 0.02 | 0.03 | 0.04 | ·········· | 0.96 | 0.97 | 0.98 | 0.99 |

$$\frac{1}{100} \quad \frac{2}{100} \quad \frac{3}{100} \quad \frac{4}{100} \quad \cdots\cdots \quad \frac{96}{100} \quad \frac{97}{100} \quad \frac{98}{100} \quad \frac{99}{100}$$

We can use a place value table to see how a number can be divided by 100.

$$1 \div 100 = 0.01$$

Units	·	tenths	hundredths
1	·	0	0
0	·	0	1

If you divide a units number by 100, the number moves two places to the right.

---

1 Write as a fraction and also as a decimal the number of hundredths that are coloured.

fraction: $\dfrac{}{100}$ $\qquad$ $\dfrac{}{100}$ $\qquad$ $\dfrac{}{100}$

decimal: [   ] $\qquad$ [   ] $\qquad$ [   ]

2 Write the missing hundredths on the number lines.

$\frac{16}{100}$ $\frac{17}{100}$ ☐ ☐ ☐ ☐

☐ ☐ 0.76 0.77 ☐ ☐

☐ ☐ ☐ ☐ ☐ 0.99

3 Divide the numbers by 100 using the place value rule. Complete the last row.

Units	·	tenths	hundredths
4	·	0	0

Units	·	tenths	hundredths
8	·	0	0

Units	·	tenths	hundredths
5	·	0	0

Units	·	tenths	hundredths
2	·	0	0

Tens	Units	·	tenths	hundredths
1	3	·	0	0

Tens	Units	·	tenths	hundredths
3	7	·	0	0

If you feel your child really understands hundredths, you could introduce percentages. 'Cent' means 100, so a percentage means 'part of one hundred'.

Decimal numbers and rounding

We already know that we can round whole numbers to the nearest ten, hundred or beyond. For example, 32 rounds **down** to 30, 176 rounds **up** to 200.

We can also round **decimal numbers** to the nearest whole number.

When a decimal number ends in a number **below 0.5** we round **down** to the nearest whole number. If it ends in **0.5 or above**, we round **up**.

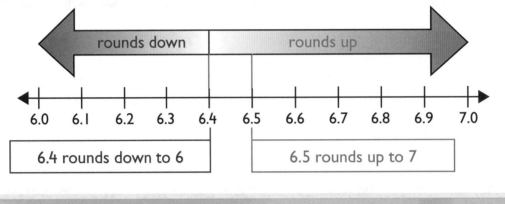

6.4 rounds down to 6

6.5 rounds up to 7

1 In each box, draw an arrow pointing down for those decimals that round down to the nearest whole number and an arrow pointing up for those that round up.
The first one has been done for you.

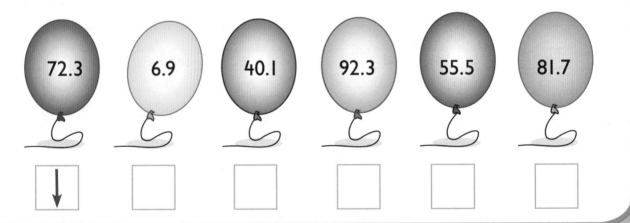

72.3   6.9   40.1   92.3   55.5   81.7

2 Complete the following sentences, rounding to the nearest whole number.

9.1 rounds down to _____   9.6 rounds up to _____

14.3 rounds down to _____   14.7 rounds up to _____

63.4 rounds down to _____   63.9 rounds up to _____

3 Round these decimals to the nearest whole number. Draw an arrow to show if the number rounds up or down and write your answer in the correct box.
The first one has been done for you.

Rounds down   Rounds up

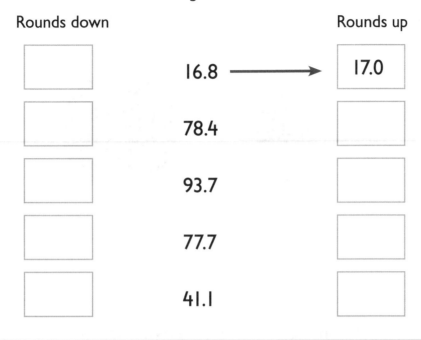

16.8 ⟶ 17.0

78.4

93.7

77.7

41.1

4 Round the numbers on the keys. Draw a line between each key and the door that it matches.

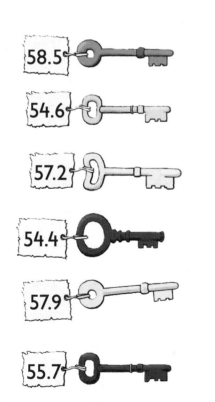

58.5

54.6

57.2

54.4

57.9

55.7

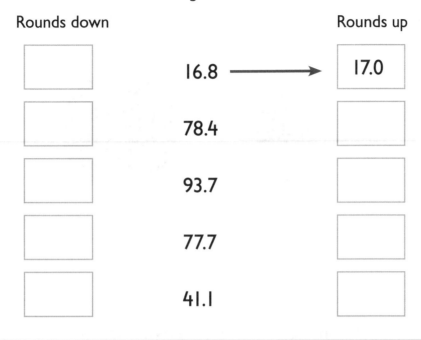

57

54

56

59

55

58

Adding fractions

When you add fractions with the **same denominators**, you only need to add the **numerators**.

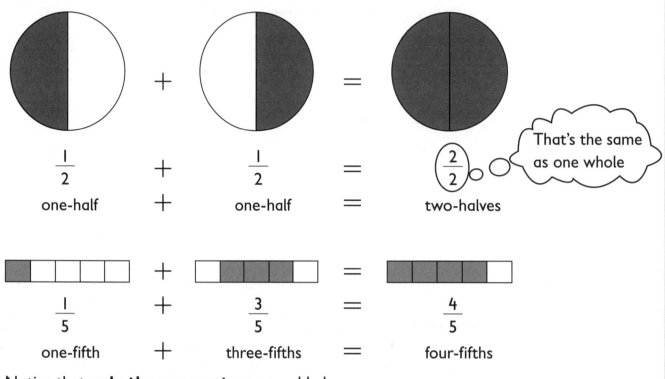

$$\frac{1}{2} + \frac{1}{2} = \frac{2}{2}$$

one-half + one-half = two-halves

That's the same as one whole

$$\frac{1}{5} + \frac{3}{5} = \frac{4}{5}$$

one-fifth + three-fifths = four-fifths

Notice that **only the numerators** are added.

1 Add the coloured parts of these shapes by colouring in the final shape.

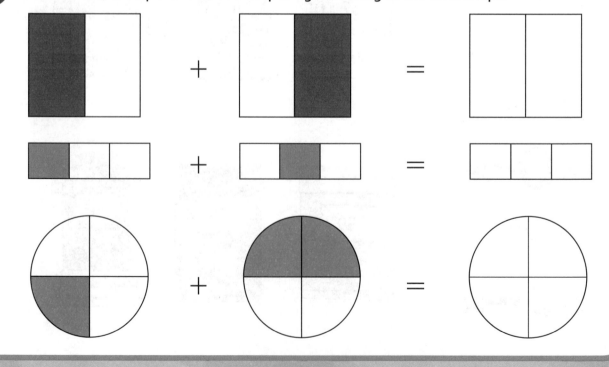

2 Add these fractions. Use the circles to help you. The first fraction has been coloured in.

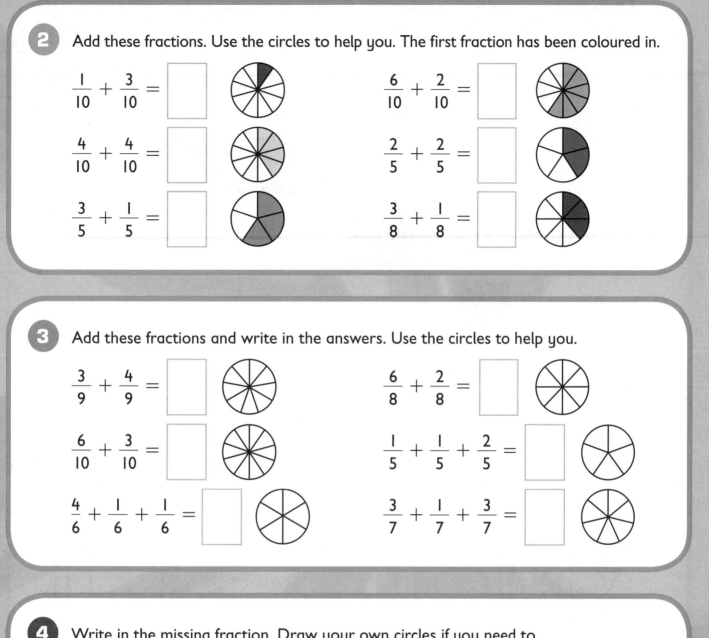

$$\frac{1}{10} + \frac{3}{10} = \boxed{}$$

$$\frac{6}{10} + \frac{2}{10} = \boxed{}$$

$$\frac{4}{10} + \frac{4}{10} = \boxed{}$$

$$\frac{2}{5} + \frac{2}{5} = \boxed{}$$

$$\frac{3}{5} + \frac{1}{5} = \boxed{}$$

$$\frac{3}{8} + \frac{1}{8} = \boxed{}$$

3 Add these fractions and write in the answers. Use the circles to help you.

$$\frac{3}{9} + \frac{4}{9} = \boxed{}$$

$$\frac{6}{8} + \frac{2}{8} = \boxed{}$$

$$\frac{6}{10} + \frac{3}{10} = \boxed{}$$

$$\frac{1}{5} + \frac{1}{5} + \frac{2}{5} = \boxed{}$$

$$\frac{4}{6} + \frac{1}{6} + \frac{1}{6} = \boxed{}$$

$$\frac{3}{7} + \frac{1}{7} + \frac{3}{7} = \boxed{}$$

4 Write in the missing fraction. Draw your own circles if you need to.

$$\frac{2}{6} + \boxed{} = \frac{5}{6}$$

$$\frac{6}{10} + \boxed{} = \frac{7}{10}$$

$$\boxed{} + \frac{2}{7} = \frac{6}{7}$$

$$\boxed{} + \frac{3}{9} = \frac{8}{9}$$

Now, see if you can balance this number statement by adding two fractions.

$$\frac{2}{6} + \boxed{} = \frac{4}{6} + \boxed{}$$

Subtracting fractions

When you subtract fractions with the **same denominators**, you only need to subtract the numerators.

You have a whole bar of chocolate with **10** pieces.

You give **6** pieces to your friend.

You have **4** pieces left.

We can write this as a number sentence $\dfrac{10}{10} - \dfrac{6}{10} = \dfrac{4}{10}$

---

1 Look at the first jug and how much tipped out. Write this as a number sentence. Then draw a line on the second jug to show how much would be left.

Three-tenths of a litre are tipped out

$$\frac{6}{10} - \boxed{} = \boxed{}$$

Five-tenths of a litre are tipped out

$$\frac{8}{10} - \boxed{} = \boxed{}$$

Six-tenths of a litre are tipped out

$$\frac{10}{10} - \boxed{} = \boxed{}$$

2 Subtract these fractions and write in the answers.

$$\frac{3}{4} - \frac{1}{4} = \boxed{}$$

$$\frac{9}{12} - \frac{6}{12} = \boxed{}$$

$$\frac{3}{5} - \frac{2}{5} = \boxed{}$$

$$\frac{8}{8} - \frac{3}{8} = \boxed{}$$

$$\frac{8}{8} - \frac{4}{8} = \boxed{}$$

$$\frac{12}{16} - \frac{9}{16} = \boxed{}$$

3 Write the number sentence for each of these problems.

A cake is cut into twelve equal pieces. You eat four of them. How much is left?

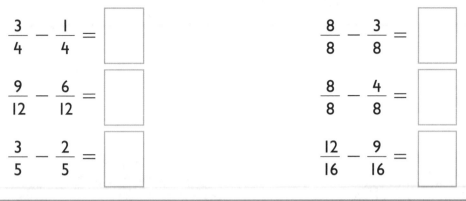

$$\frac{12}{12} - \boxed{} = \boxed{}$$

There are sixth-eighths of a pizza left. You eat two of these. How much is left now?

$$\frac{6}{8} - \boxed{} = \boxed{}$$

A watermelon is cut into quarters. One-quarter falls on the floor. How much is left?

$$\frac{4}{4} - \boxed{} = \boxed{}$$

Encourage your child to link the fraction to an amount. It doesn't matter if the fraction is about something solid, like pebbles or liquid, like juice. You could use the same paper strip idea to demonstrate subtracting fractions as for adding.

Equivalent fractions

Equivalent fractions have the same value even though they look different.

$\frac{1}{2}$ and $\frac{2}{4}$ are **equivalent fractions**.

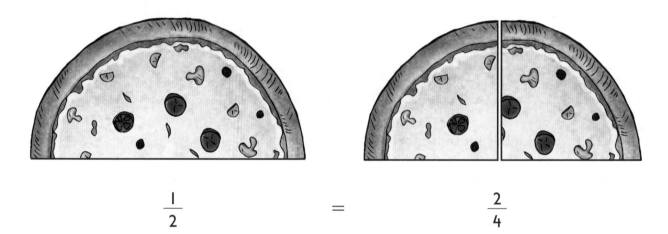

$$\frac{1}{2} \qquad = \qquad \frac{2}{4}$$

They have different denominators and numerators but their value is the same.

---

1 Colour the correct number of sections.
In each box, write the fraction of the circle you have coloured.

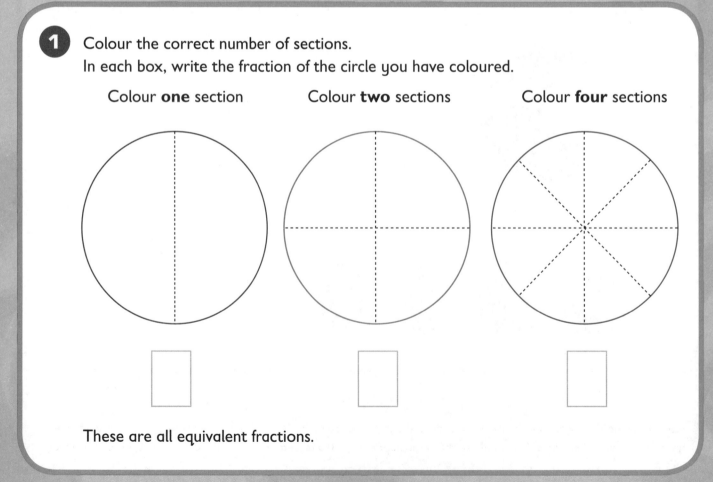

Colour **one** section   Colour **two** sections   Colour **four** sections

These are all equivalent fractions.

2 Look at the instructions for each circle. Colour the correct number of sections. In each box, write the fraction of each circle you have coloured.

Colour **one** section   Colour **two** sections   Colour **four** sections

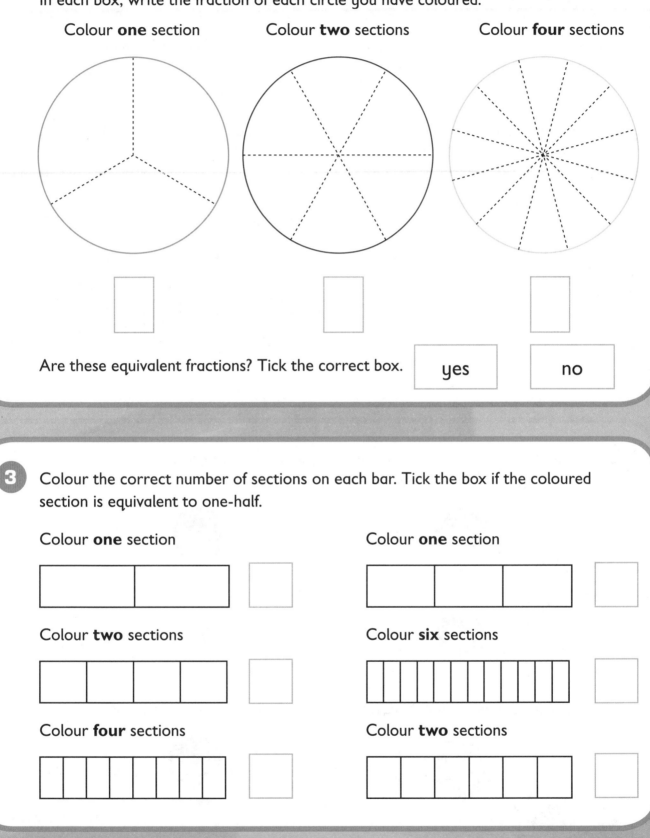

Are these equivalent fractions? Tick the correct box.

| yes | no |

3 Colour the correct number of sections on each bar. Tick the box if the coloured section is equivalent to one-half.

Colour **one** section   Colour **one** section

Colour **two** sections   Colour **six** sections

Colour **four** sections   Colour **two** sections

A fraction wall is a great way of seeing how different fractions are equivalent to each other. You can easily make one by using strips of paper that are all the same length, and fold them into smaller pieces.

More equivalent fractions

Equivalent fractions have equal value even though they look different. If we double both the denominator and also the numerator of one fraction, we create an equivalent fraction.

Working **down** a fraction wall can help us to understand equivalent fractions. Look at the left half of the fraction wall and you can see that

$$\frac{1}{2} = \frac{2}{4} = \frac{4}{8} = \frac{8}{16} = \frac{16}{32}$$

Working **up** a fraction wall can help us to **simplify** fractions. This is where the fraction is written as the smallest possible numbers. You can see that

$$\frac{16}{32}, \frac{8}{16}, \frac{4}{8}, \frac{2}{4} \text{ all simplify to } \frac{1}{2}$$

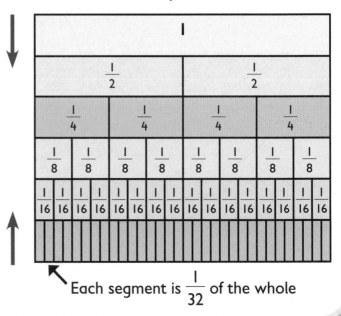

Each segment is $\frac{1}{32}$ of the whole

---

1 Three of these children have eaten the same amount of pizza. Tick the child who has eaten more than their friends.

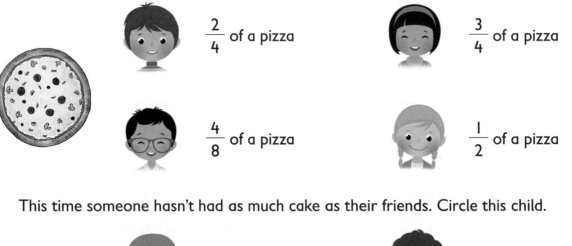

$\frac{2}{4}$ of a pizza

$\frac{3}{4}$ of a pizza

$\frac{4}{8}$ of a pizza

$\frac{1}{2}$ of a pizza

This time someone hasn't had as much cake as their friends. Circle this child.

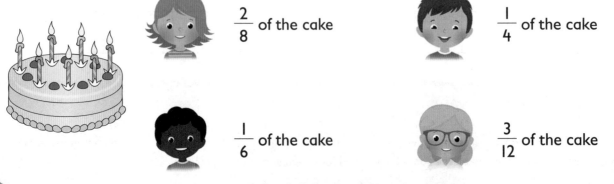

$\frac{2}{8}$ of the cake

$\frac{1}{4}$ of the cake

$\frac{1}{6}$ of the cake

$\frac{3}{12}$ of the cake

2 Complete these equivalent fractions by doubling the numerator and the denominator each time.

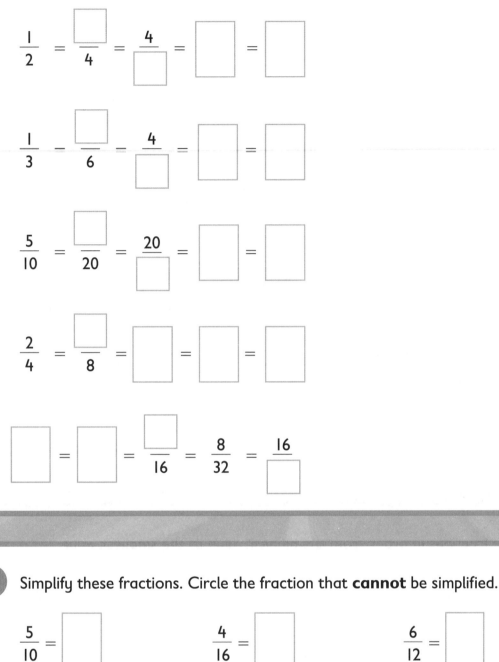

$$\frac{1}{2} = \frac{\square}{4} = \frac{4}{\square} = \square = \square$$

$$\frac{1}{3} = \frac{\square}{6} = \frac{4}{\square} = \square = \square$$

$$\frac{5}{10} = \frac{\square}{20} = \frac{20}{\square} = \square = \square$$

$$\frac{2}{4} = \frac{\square}{8} = \square = \square = \square$$

$$\square = \square = \frac{\square}{16} = \frac{8}{32} = \frac{16}{\square}$$

3 Simplify these fractions. Circle the fraction that **cannot** be simplified.

$$\frac{5}{10} = \square \qquad \frac{4}{16} = \square \qquad \frac{6}{12} = \square$$

$$\frac{2}{6} = \square \qquad \frac{3}{4} = \square \qquad \frac{8}{12} = \square$$

Ordering fractions by size

Ordering unit fractions by size can be done by comparing the **denominators**.

$\dfrac{1}{2}$ is larger than $\dfrac{1}{8}$

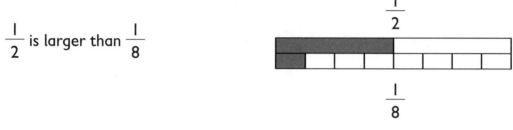

$\dfrac{1}{2}$

$\dfrac{1}{8}$

The way we order fractions with the **same** denominator but **different** numerators is reversed – the bigger the numerator, the higher the value of the fraction.

$\dfrac{5}{8}$ is larger than $\dfrac{1}{8}$

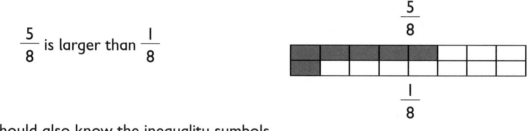

$\dfrac{5}{8}$

$\dfrac{1}{8}$

You should also know the inequality symbols,

> means **greater than** and < means **less than**.

$\dfrac{1}{2} > \dfrac{1}{8}$

one-half is greater than one-eighth

$\dfrac{1}{8} < \dfrac{5}{8}$

one-eighth is less than five-eighths

---

1 Write these fractions in order, from the highest value to the smallest value.

$\dfrac{4}{8}$ $\qquad$ $\dfrac{1}{8}$ $\qquad$ $\dfrac{7}{8}$ $\qquad$ $\dfrac{3}{8}$ $\qquad$ $\dfrac{5}{8}$ $\qquad$ $\dfrac{6}{8}$

highest $\qquad\qquad\qquad\qquad\qquad\qquad\qquad\qquad\qquad\qquad\qquad$ smallest

$\dfrac{2}{12}$ $\qquad$ $\dfrac{9}{12}$ $\qquad$ $\dfrac{4}{12}$ $\qquad$ $\dfrac{11}{12}$ $\qquad$ $\dfrac{1}{12}$ $\qquad$ $\dfrac{6}{12}$

highest $\qquad\qquad\qquad\qquad\qquad\qquad\qquad\qquad\qquad\qquad\qquad$ smallest

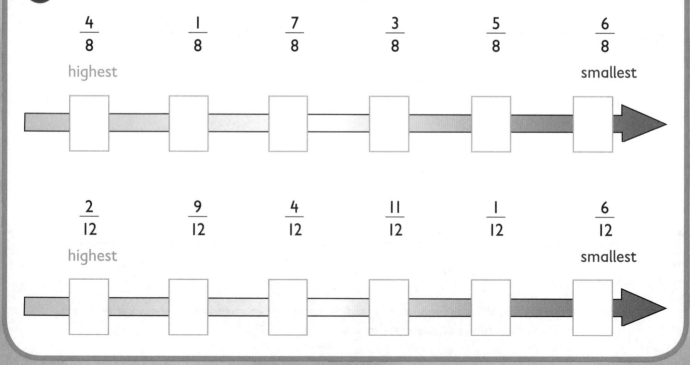

2 Colour one section of each circle. Then, in the answer box, write the fraction of each circle you have coloured.

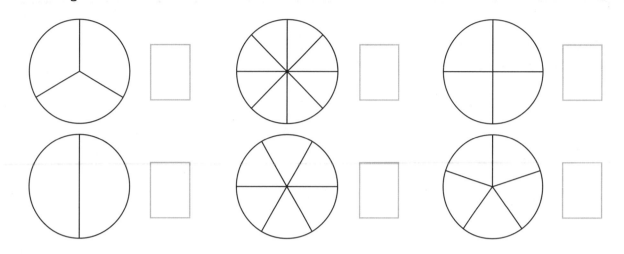

Rewrite these fractions in order, starting with the highest value.

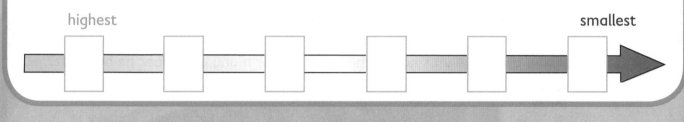

highest   smallest

3 Write < > or = between these pairs of fractions. Use the fraction wall on page 18 to help or try drawing one of your own.

| $\frac{6}{8}$ | $\frac{3}{8}$ | | $\frac{1}{3}$ | $\frac{1}{5}$ | | $\frac{1}{12}$ | $\frac{1}{6}$ |

| $\frac{3}{4}$ | $\frac{2}{4}$ | | $\frac{8}{10}$ | $\frac{6}{10}$ | | $\frac{1}{16}$ | $\frac{1}{5}$ |

| $\frac{4}{8}$ | $\frac{1}{2}$ | | $\frac{5}{10}$ | $\frac{1}{4}$ | | $\frac{5}{10}$ | $\frac{1}{2}$ |

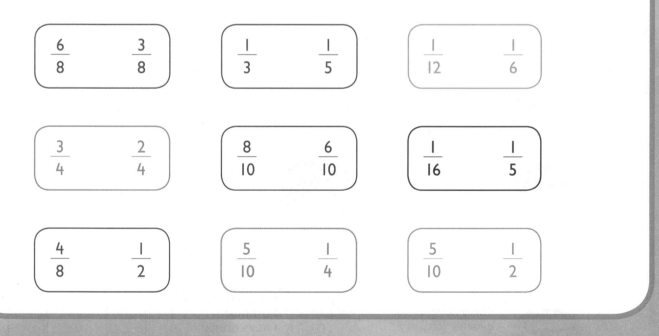

Finding a fraction of a quantity

Quantity means the amount or number of something. You may not have the whole quantity but you may have a fraction of the whole.

All the pears (whole) $\frac{1}{2}$ of the pears $\frac{1}{4}$ of the pears

1   Tom is baking cakes. He uses half of these eggs. How many eggs does he have left?

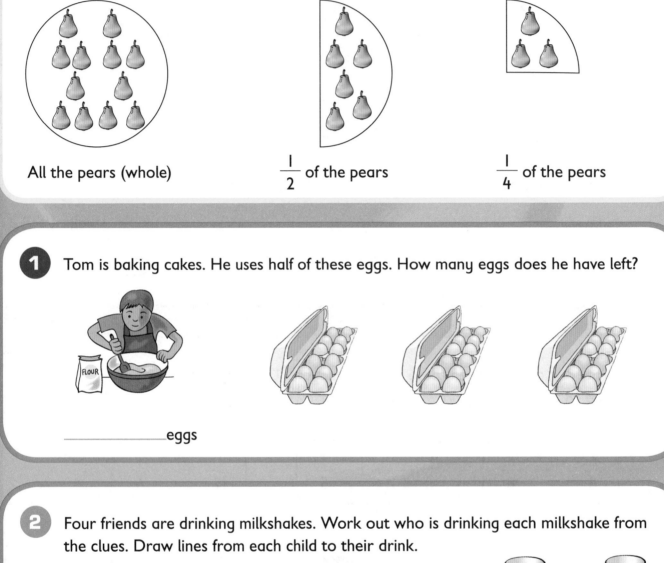

_____ eggs

2   Four friends are drinking milkshakes. Work out who is drinking each milkshake from the clues. Draw lines from each child to their drink.

Josh: I have only drunk a quarter of mine.

Dane: My milkshake is $\frac{2}{3}$ empty.

Morgan: I only have a quarter left!

Nate: I can't remember how much I have drunk!

3 A quarter of the fruits from each of these trees is picked. The number of fruits picked is on each basket. In the boxes, write how much fruit is left on each tree.
Draw apples on the tree if it helps.

4 A family want to get a dog. They visit the kennel and look at three dogs. They want a dog that won't cost too much to feed. How much food does each dog eat each day?

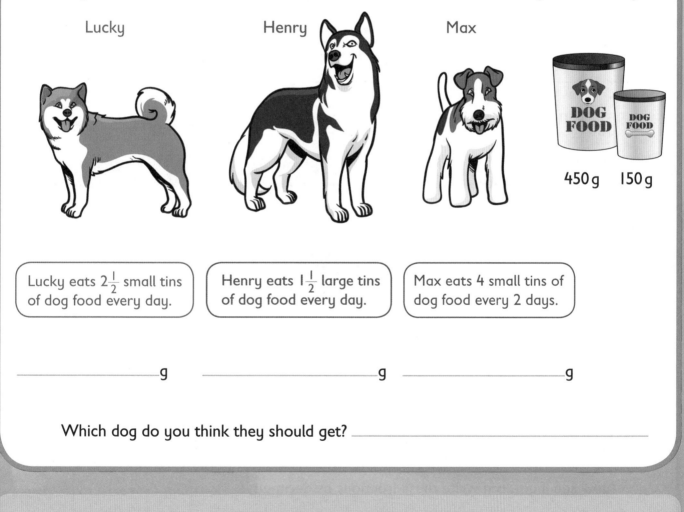

Lucky

Henry

Max

450 g   150 g

| Lucky eats $2\frac{1}{2}$ small tins of dog food every day. | Henry eats $1\frac{1}{2}$ large tins of dog food every day. | Max eats 4 small tins of dog food every 2 days. |

_____ g   _____ g   _____ g

Which dog do you think they should get? _____

A common mistake children make is to think of the number one as 'the whole'. This is because the first fractions they learn are smaller than one. Try to get your child used to the idea that the whole doesn't always refer to one of something. It could be 'one whole' box of raisins, or 'one whole' bunch of balloons.

Matching equivalent fractions and decimals

Decimals are fractions written in a different way.
Place value helps us to understand decimals.

Units	·	tenths	hundredths
0	·	1	0
0	·	1	5

Hundredths can help us to write decimal numbers. Imagine one whole square divided into 100 tiny squares. Each square is one-hundredth. The diagrams below show equivalent fractions and decimals.

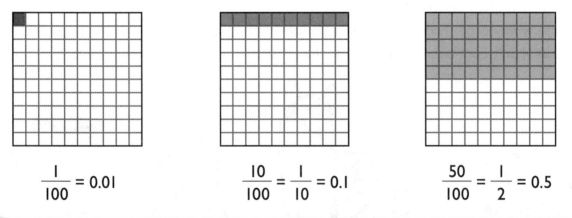

$$\frac{1}{100} = 0.01 \qquad \frac{10}{100} = \frac{1}{10} = 0.1 \qquad \frac{50}{100} = \frac{1}{2} = 0.5$$

---

1 Write the correct decimals on the answer lines.

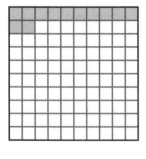
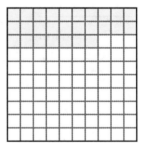

$$\frac{12}{100} = \underline{\hspace{3cm}} \qquad \frac{34}{100} = \underline{\hspace{3cm}} \qquad \frac{27}{100} = \underline{\hspace{3cm}}$$

Now see if you can write these fractions as decimals.

$$\frac{99}{100} = \underline{\hspace{3cm}} \qquad \frac{50}{100} = \underline{\hspace{3cm}} \qquad \frac{8}{100} = \underline{\hspace{3cm}}$$

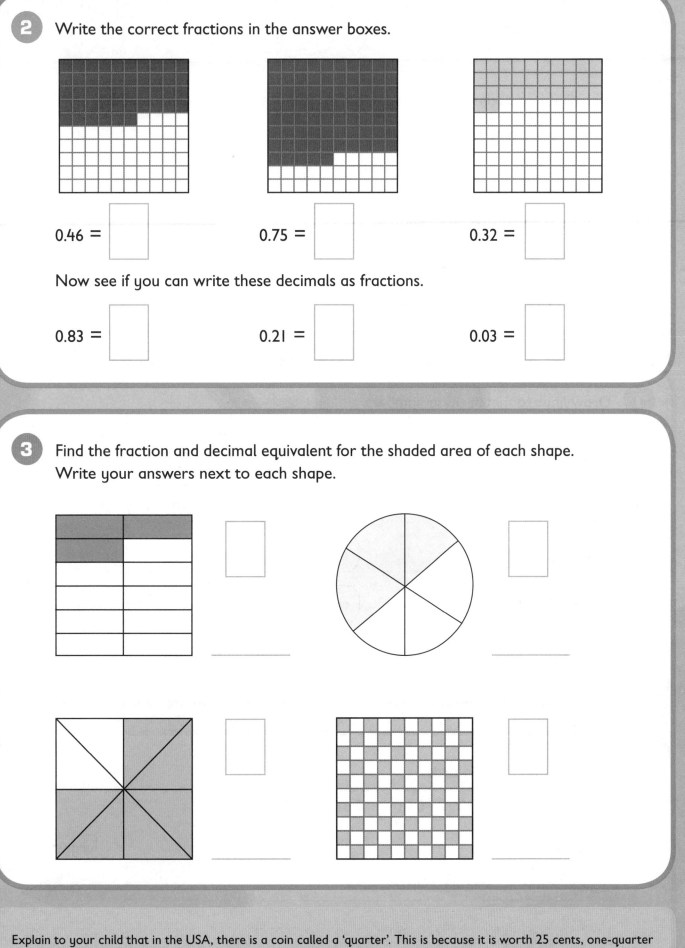

2 Write the correct fractions in the answer boxes.

0.46 = ☐   0.75 = ☐   0.32 = ☐

Now see if you can write these decimals as fractions.

0.83 = ☐   0.21 = ☐   0.03 = ☐

3 Find the fraction and decimal equivalent for the shaded area of each shape.
Write your answers next to each shape.

Explain to your child that in the USA, there is a coin called a 'quarter'. This is because it is worth 25 cents, one-quarter of 1 dollar (100 cents).

Putting decimal numbers in order

Read each digit in a decimal separately and think about the place value carefully.

Units	·	tenths	hundredths
3	·	4	2
3	·	6	0

3.42 = three and forty-two hundredths

3.60 = three and sixty hundredths

A common mistake is to read 3.42 as 'three point forty two'. This makes it appear bigger than 'three point six'.

1 Draw lines to match these numbers.

3.74   eight and four hundredths

8.04   eight and ninety hundredths

8.90   eleven and twenty-three hundredths

6.80   three and seventy-four hundredths

11.23   six and eighty hundredths

2 Write the numbers in order.

| 0.47 | 0.1 | 0.13 | 0.01 | 0.31 | 0.74 |

smallest   largest

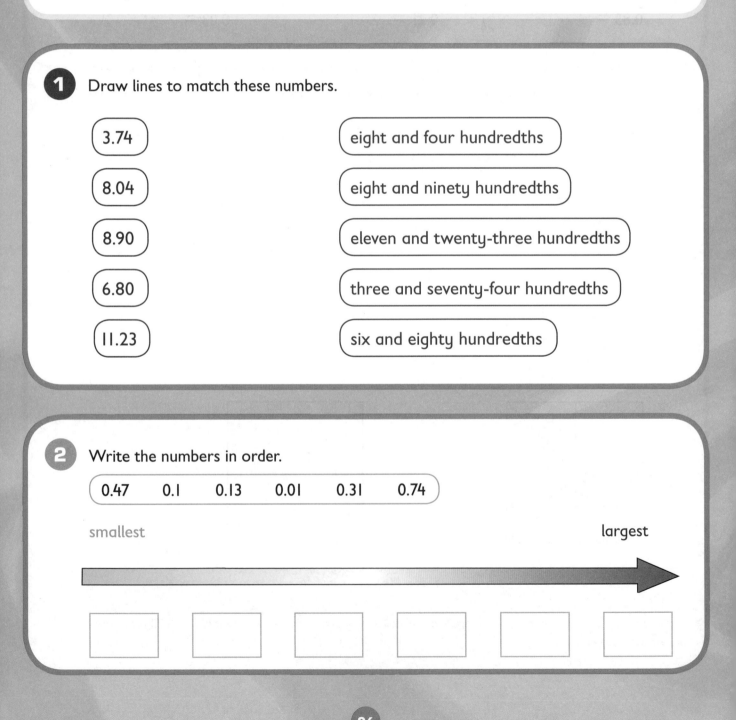

3 Draw a circle around the correct inequality symbol.

32.1	<	>	32.2		12.6	<	>	12.4
76.5	<	>	76.2		29.19	<	>	29.16
84.42	<	>	84.43		135.27	<	>	135.26

4 Write the numbers in order.

| 23.46 | 2.34 | 2.04 | 234.6 | 234.64 | 23.64 |

smallest   largest

5 Write a shopping list putting the items in order from the most expensive to the cheapest.

£2.07 Olives

£1.40 Pickles

£2.70 Marmalade

£1.04 Strawberry Jam

Shopping list

1 _____

2 _____

3 _____

4 _____

You could explain that putting decimal numbers in order is similar to putting words in alphabetical order. There is a system of looking at the first letter, then the next, and so on. It is the same with numbers.

Problem solving and reasoning

These pages are all about reasoning and thinking logically. Use your knowledge of fractions to complete the problems and puzzles.

1 Can you follow the clues to colour the shape? What fraction is left uncoloured?

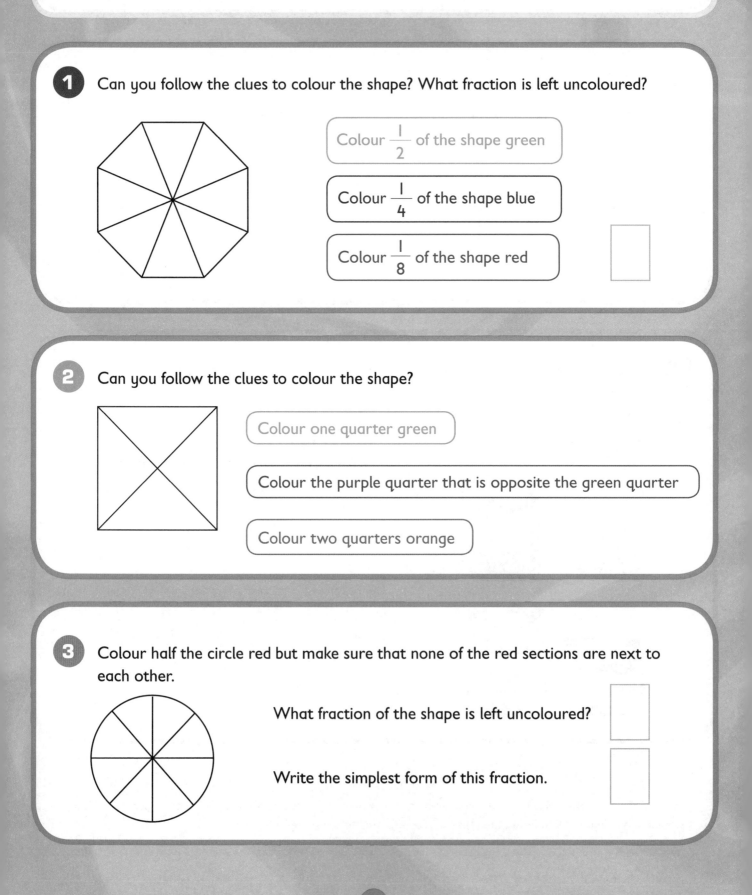

Colour $\dfrac{1}{2}$ of the shape green

Colour $\dfrac{1}{4}$ of the shape blue

Colour $\dfrac{1}{8}$ of the shape red

2 Can you follow the clues to colour the shape?

Colour one quarter green

Colour the purple quarter that is opposite the green quarter

Colour two quarters orange

3 Colour half the circle red but make sure that none of the red sections are next to each other.

What fraction of the shape is left uncoloured?

Write the simplest form of this fraction.

4 Two children are playing a game. They pick a card with a fraction on it and colour in the correct number of squares on their grid. The player who colours in all of their grid correctly first is the winner. Here is their game so far:

Player A  Player B

Who is winning? How do you know? _____

Player A's next card is $\dfrac{1}{2}$. Player B's next card is $\dfrac{3}{8}$. Add these fractions to their grids.

What is the score now? Player A _____ Player B _____

What fraction is needed for each player to finish? Player A ☐ Player B ☐

5 Three children are playing a turning game. Here is what they are saying:

If you turn one and a half times round, you face the same way as if you turned halfway round.

You would also face the same way if you turned one quarter, and then another quarter the same way.

That doesn't sound the same.

Do you agree with the girl in the green shirt? Why?
You could try doing the turns yourself to see if they are correct.

_____

Money problems

When you go shopping, you may see different offers like ' $\frac{1}{2}$ price sale' or ' $\frac{1}{3}$ off'. Use your knowledge of fractions and decimals to get a bargain!

1 Can you use your knowledge of fractions and decimals to find out the new sale prices? Write your answers in the boxes.

£20.50

£42.50

£35

2 There are two pairs of trainers in the sale. Write the sale price for each pair of trainers in each answer box.
Tick the pair of trainers that offers a greater saving.

$\frac{1}{3}$ off !

$\frac{1}{4}$ off !

£120

£88

3 Look at the bus ticket prices for a trip to the seaside.

Adult: £1.50

Child: half price

Child under 4: free

Over 60s: $\frac{1}{3}$ off adult price

How much will it cost for this family to get the bus to the seaside?
Use the box below for working out and write your answer in the answer box.

| Mum   A child aged 6   A child aged 10   A child aged $2\frac{1}{2}$ |

How much will it cost for this family to get to the seaside?
Use the box below for working out and write your answer in the answer box.

| Grandma and Grandpa, both over 60   Their grandson, aged $3\frac{1}{2}$ |

The bus driver found a ticket costing £4.50.
How many adults and children could this ticket be for?
There are several different answers. See how many you can find.

Answers

Page 4

What are fractions and decimals?

1 fractions: $\frac{1}{7}, \frac{1}{9}, \frac{3}{4}, \frac{8}{10}$

decimals: 6.8, 12.3, 19.8, 31.5

2 3.5, 61.9, 55.8, 70.5, 28.6

Page 5

3 Eight units, six tenths
Seven tens, zero units, seven tenths
One ten, six units, two tenths

4 $\frac{2}{5}$ non-unit fraction

$\frac{4}{12}$ non-unit fraction

$\frac{7}{8}$ non-unit fraction

$\frac{1}{2}$ unit fraction

5 $\frac{1}{7}$ $\frac{1}{8}$

Could be $\frac{2}{7}, \frac{3}{7}, \frac{4}{7}, \frac{5}{7}, \frac{6}{7}, \frac{7}{7}$

Could be $\frac{2}{6}, \frac{3}{6}, \frac{4}{6}, \frac{5}{6}, \frac{6}{6}$

Could be $\frac{2}{5}, \frac{3}{5}, \frac{4}{5}, \frac{5}{5}$

$\frac{1}{5}$

Page 6

Tenths

1 0.1, 0.2, 0.3, 0.4, 0.5, 0.6, 0.7, 0.8, 0.9, 1.0

$\frac{10}{10}, \frac{9}{10}, \frac{8}{10}, \frac{7}{10}, \frac{6}{10}, \frac{5}{10}, \frac{4}{10}, \frac{3}{10},$

$\frac{2}{10}, \frac{1}{10}$

Page 7

2 $\frac{1}{10}$ 0.1 $\frac{3}{10}$ 0.3

$\frac{4}{10}$ 0.4 $\frac{7}{10}$ 0.7

$\frac{9}{10}$ 0.9

0.8 is the odd one out.

3 1 token coloured
6 sweets coloured

4 0.6, 0.5, 0.8, 0.2, 0.3, 0.1, 0.7, 0.4

Page 8

Hundredths

1 $\frac{9}{100}$ 0.09, $\frac{40}{100}$ 0.40, $\frac{90}{100}$ 0.90

Page 9

2 $\frac{16}{100}, \frac{17}{100}, \frac{18}{100}, \frac{19}{100}, \frac{20}{100}, \frac{21}{100}$
0.74, 0.75, 0.76, 0.77, 0.78, 0.79
0.94, 0.95, 0.96, 0.97, 0.98, 0.99

3 0.04, 0.08, 0.05, 0.02, 0.13, 0.37

Page 10

Decimal numbers and rounding

1 Arrow down: 40.1, 92.3
Arrow up: 6.9, 55.5, 81.7

2 Rounds down: 9.0, 14.0, 63.0
Rounds up: 10.0, 15.0, 64.0

Page 11

3 Left: 78.0, 41.0 Right: 94.0, 78.0

4 58.5 59   54.6 55
57.2 57   54.4 54
57.9 58   55.7 56

Page 12

Adding fractions

1

Page 13

2 $\frac{1}{10} + \frac{3}{10} = \frac{4}{10}$ $\frac{6}{10} + \frac{2}{10} = \frac{8}{10}$

$\frac{4}{10} + \frac{4}{10} = \frac{8}{10}$ $\frac{2}{5} + \frac{2}{5} = \frac{4}{5}$

$\frac{3}{5} + \frac{1}{5} = \frac{4}{5}$ $\frac{3}{8} + \frac{1}{8} = \frac{4}{8}$

3 $\frac{3}{9} + \frac{4}{9} = \frac{7}{9}$ $\frac{6}{8} + \frac{2}{8} = \frac{8}{8}$

$\frac{6}{10} + \frac{3}{10} = \frac{9}{10}$ $\frac{1}{5} + \frac{1}{5} + \frac{2}{5} = \frac{4}{5}$

$\frac{4}{6} + \frac{1}{6} + \frac{1}{6} = \frac{6}{6}$

$\frac{3}{7} + \frac{1}{7} + \frac{3}{7} = \frac{7}{7}$

4 $\frac{2}{6} + \frac{3}{6} = \frac{5}{6}$ $\frac{6}{10} + \frac{1}{10} = \frac{7}{10}$

$\frac{4}{7} + \frac{2}{7} = \frac{6}{7}$ $\frac{5}{9} + \frac{3}{9} = \frac{8}{9}$

There are many answers e.g.

$\frac{2}{6} + \frac{3}{6} = \frac{4}{6} + \frac{1}{6}$ or

$\frac{2}{6} + \frac{4}{6} = \frac{4}{6} + \frac{2}{6}$

Page 14

Subtracting fractions

1 $\frac{6}{10} - \frac{3}{10} = \frac{3}{10}$ A line is drawn at $\frac{3}{10}$ on the jug.

$\frac{8}{10} - \frac{5}{10} = \frac{3}{10}$ A line is drawn at $\frac{3}{10}$ on the jug.

$\frac{10}{10} - \frac{6}{10} = \frac{4}{10}$ A line is drawn at $\frac{4}{10}$ on the jug.

Page 15

2 $\frac{3}{4} - \frac{1}{4} = \frac{2}{4}$ $\frac{8}{8} - \frac{3}{8} = \frac{5}{8}$

$\frac{9}{12} - \frac{6}{12} = \frac{3}{12}$ $\frac{8}{8} - \frac{4}{8} = \frac{4}{8}$

$\frac{3}{5} - \frac{2}{5} = \frac{1}{5}$ $\frac{12}{16} - \frac{9}{16} = \frac{3}{16}$

3 $\frac{12}{12} - \frac{4}{12} = \frac{8}{12}$

$\frac{6}{8} - \frac{2}{8} = \frac{4}{8}$

$\frac{4}{4} - \frac{1}{4} = \frac{3}{4}$

Page 16

Equivalent fractions

1 $\frac{1}{2}$ $\frac{2}{4}$ $\frac{4}{8}$

Page 17

2 $\frac{1}{3}$ $\frac{2}{6}$ $\frac{4}{12}$

Yes they are equivalent.

3 There are many different combinations of colouring.

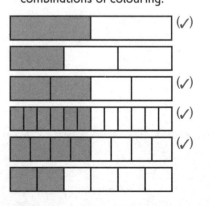